Orbital Debris, the problem and the mitigation

Patrick H. Stakem

2018

Number 17 in the Space series.

Table of Contents

Introduction..3
Introduction..3
The Author..3
A note on Units..4
The debris problem..5
    Space Debris...7
    Zombie Sats...8
    Satellite on-orbit collision...9
    Booster debris..9
    The graveyard orbit...10
What did we do with the Apollo hardware?..11
SMM...13
Space Station Mir...14
Shuttle..15
ISS..16
The LDEF Mission..17
Saturn-Pegasus..18
Skylab...24
NASA Orbital Debris Program Office...25
Tracking Junk, NORAD...28
Wrap-up..30
Bibliography...32
Resources..40
Glossary...46
If you enjoyed this book, you may have an interest in some of the author's other books:..50

Introduction

This book covers the topic of Orbital debris, what it is, where it comes from, what problems it introduces, and how to deal with it. . Putting a communications satellite in synchronous orbit will set you back 100's of millions of dollars. Once on orbit, you hope it survived the launch environment, and operates correctly. You further hope it works at least for its design lifetime, and as long as possible. This approach, based on good engineering design practices, lessons learned, and hope, it the equivalent of buying a new Tesla with non-rechargeable batteries, and driving it until it stops. Then buying a new one. Regardless of what you were told, there is no satellite fairy with a magic wand. There will be some serious trash-talk here. This third edition has some breaking news that happened as the second edition was published.

The Author

The author spent his career in support of numerous NASA spaceflight missions. He has taught for the graduate Computer Science Department of Loyola University in Maryland, and for Capitol Technology University. He currently teaches for the Johns Hopkins University, Whiting School of Engineering, Engineering for Professionals Program,

He spent 42 years as a NASA contractor, at all of the NASA sites. He began his aerospace career at Fairchild

Industries, and overlapped with Dr. Wernher von Braun, so, technically, he was a member of the von Braun rocket team. He has received the NASA Shuttle Program Managers Commendation Award, two NASA Group Achievement Awards, Certificate of Appreciation from the NASA Earth Science Technology Office, and the NASA Apollo-Soyuz Test Program Award. He did extensive work on the Flight Telerobotic Servicer Mission. He served on AIAA's Committee on Standards for Automation and Robotics for TransOrbital, Lunar and Mars Base. He worked on the Solar Maximum Repair Mission, and briefly on the Hubble Space Telescope repair missions.

A note on Units

I am fairly conversant in both English and Metric units (what is the metric equivalent of furlongs per fortnight?). Metric (SI) is mandated for NASA usage now, for interchangeability with our partner space faring nations. When a lot of the legacy flights discussed here were flown, English units were the norm. I have tried to keep the units comparable to the mission at the time. Conversions are easy enough, but units conversion is a source of error. It's not what you know about units and measurement, its how you think. And, I still think English units (even the English use Metric now), and convert in my head or on my phone.

For scientific/engineering work, the Metric system is well thought out. For artisans, the English system served well, as most units were divided by 2. Which is easy. Fold the cloth. Hopefully, when we are all taught Metric first, some one will still remember the back conversions. You just need a good slide rule....

The debris problem

Orbital debris can come from many sources. It can be naturally occurring rocks and captured meteoroids. It can come from satellites collisions and explosions. Liquid fuel, from a dead satellite or leakage, freezes in space and provides yet more debris problems. There have been 5 known collisions of satellites in orbit so far.

All of the junk, down to the size of bolts, is tracked by the U.S. Air Force. They know of 18,000 objects in orbit, of which 1,400 are operational satellites. A good job for a robotic servicer in LEO would be to collect the trash, put it in a canister, and kick it off to re-enter and burn in the atmosphere. There are estimated to be 170 million chunks, smaller than a centimeter, any of which can ruin your mission.

Known space debris includes Astronaut Ed White's outer glove, lost on his space walk; Michael Collins's camera from Gemini-10; a wrench, pair of pliers, and a tooth brush; and a complete tool bag from an STS-26 EVA, and suitsat, an old Russian Spacesuit, put out an airlock on the ISS.

In addition to our own junk, space itself provides dust particles, up to and including asteroids capable of ending all life on Earth with a collision.

For mitigation, the ISS uses Whipple shielding, named for its inventor. This uses a thin outer bumper, spaced away from the hull. The idea is, the bumper breaks up the debris so that penetration doesn't happen. It works most of the time. Think of it as the ISS's bullet proof vest. It works best with a foam material behind it.

The US is responsible for the most debris in space, followed by Russia and China.

There have been on-orbit tests of anti-satellite weapons, by the U.S., Russia, and China which resulted in a large amount of debris. In addition, failed on-orbit rendezvous usually involved a low speed collision, and more debris creation. There was the 1996 collision between the French Cerise military reconnaissance satellite and debris from the Ariane rocket. In addition, a 2009 collision between the Iridium 33 communications satellite and the derelict Russian Kosmos 2251 resulted in the destruction of both satellites, and a big mess. There was the 2013 collision between debris from the Chinese Fungyun FY-1C satellite and the Russian BLITS nano-satellite. We should also mention the 2013 collision between two CubeSats, Ecuador's NEE-01 Pegaso and Argentina's CubeBug-1, and the particles of a debris cloud around a Russian Tsyklon-3 upper stage from the launch of Kosmos 1666. A 1985 U.S. Anti-satellite test created thousands pieces of debris greater than 1 cm. Most of this reentered the atmosphere within 10 years.

In 2009, a U.S. Iridium satellite collided with a defunct Russian Cosmos satellite, producing some 2,100 pieces of debris. There is a catastrophic collision every 5-9 years.

China is responsible for the largest single space debris incident resulting from a 2007 test of anti-satellite weapons. It transformed one satellite into nearly 2,500 pieces, golf-ball size or larger. The test was conducted in an orbital zone where most near-Earth missions are located. Some satellites have had to maneuver to avoid collisions.

Space Debris

Sometimes we are out own worst enemies. Orbital debris can come from many sources. It can be naturally occurring rocks and captured meteoroids. It can come from satellites collisions and explosions. Liquid fuel, from a dead satellite or leakage, freezes in space and provides yet more debris problems. There have been 5 known collisions of satellites in orbit so far.

All of the junk, down to the size of bolts, is tracked by the U.S. Air Force. They know of 18,000 objects in orbit, of which 1,400 are operational satellites. A good job for a robotic servicer in LEO would be to collect the trash, put it in a canister, and kick it off to re-enter and burn in the atmosphere. There are estimated to be 170 million chunks, smaller than a centimeter, any of which can ruin your mission.

NASA says that at least one piece of space debris falls to Earth daily, and has for the past 50 years. There have been injuries, such as the Japanese fishermen hit in 1969. A lady in Oklahoma was hit by a piece of debris in 1997 but was not injured. This turned out to be a pieces of a Delta-II rocket propellant tank.

Currently, more than 27000 pieces of orbital debris, or "space junk," are tracked by the Department of Defense's Global Space Surveillance Network.

Just as I was writing this, a chunk of the defunct Iridium-70 satellite hit California.

Zombie Sats

Zombie sats are non-functioning satellites in orbit. They may have experienced a failure, and are no longer functional. They remain in the same slowly-decaying orbit, however. The Intelsat Galaxy-15 is an example. It was in geostationary orbit when the ground lost control, and it began to drift. There was a potential of collision with other, operating satellites. Later, control was recovered, and it was directed back to its correct orbital position.

The first U. S. satellite, Vanguard-1, and its upper stage are still in-orbit. It was launched in 1958. The Soviet-era RORSAT, with its BES-5 nuclear reactor, is still in orbit. In 2015, the USAF Defense Meteorological Satellite (Military equivalent of the GOES weather satellite) exploded in orbit, creating some 150 chunks of satellite in orbit.

Of the 7,000 or so satellites place into Earth orbit so far, 1,500 are still functioning. The rest are zombie-Sats.

Satellite on-orbit collision

There was a collision between two satellites occurred in February of 2009. One was a Russian Strela-class military satellite, massing 950 kilograms. The other was the commercial Iridium-33 communications satellite. What was the cause? They were in the same place at the same time. The Russian spacecraft had been deactivated, and was classified as space debris. The Iridium was operational, and was destroyed.

And, the bad news is, the collision created a thousand pieces of space debris larger than 4 inches, and many more smaller ones. In March 2012, a piece of the KOSMOS 2251 passed by the International Space Station, prompting the crew to take refuge in the attached Soyuz return craft as a precaution. The ISS frequently does obstacle avoidance maneuvers.

The SNAP-10 spacecraft was launched in 1965. It included a nuclear fission device for electrical power. It operated for 45 days before an electrical failure. It was placed in a 1,300 km orbit, where it is expected to remain for around 4,000 years. It was noticed that in 1979, the spacecraft was shedding traceable debris. It is not know if radioactive material is included.

Booster debris

After the first stage of the launcher has done its job, it usually crashes into a remote part of the ocean. The next stage may also. The final stage usually accompanies the payload into orbit. The Shuttles' boosters were dropped in the Atlantic, recovered, refurbished, and reused. The large liquid fuel tank was crashed into a remote part of the ocean. These are a particular problem, because they always have residual fuel, which decomposes.

A Chinese Long March-4 upper stage exploded in orbit in 2000, creating a vast debris cloud. A Russian booster exploded in 2007, captured in images. The debris cloud was tracked on radar. More than 1,000 fragments were identified.

In 2001, part of a U. S. upper stage of the NavStar-32 satellite crashed into the Saudi Arabian desert.

The graveyard orbit.

A graveyard orbit is a disposal area for non-operational spacecraft. This is particularly important for the geosynchronous missions, which tend to be commercial. The geo-sync graveyard orbit is a few hundred kilometers higher than the operational orbit. It is much less costly in propellant to place that defunct spacecraft in the slightly higher orbit, than to re-enter it. This doesn't solve the problem completely, but postpones it until servicing options become available.

As part of the licensing process for satellites providing communication services from geo-sync, a requirement is imposed to move the spacecraft to a graveyard orbit at the end of its operational life.

Where do old, non-operational spacecraft wind up? Mostly in the ocean. The Chinese Tiangong-1 space station will reenter to atmosphere sometime around when this book gets wrapped up. Ground controllers have lost control, so they don't know where it will come down, until shortly before it does.

The point in the Oceans farthest from land is called the pole of inaccessibility. This is the best point to aim for, if you're de-orbiting something large. It happens to be located in the South Pacific, some 1,600 miles south of the Pitcairn Islands. There are some 260 satellites on the ocean floor at that point. This will be an interesting place for future archaeologists. Among other things, the 120-ton MIR space station is there. It's entry into the water was observed by fishermen. This is also where supply modules from the ISS, loaded with trash, are sent. Sometimes, things don't quite work out. 36 tons of Salyut Space Station came down in South America.

What did we do with the Apollo hardware?

The Apollo missions to the moon consisted of the crewed Apollo capsule, the Lunar Excursion Module, and the Service Module. Only the crewed capsule and the service module returned from the moon. Only the crewed capsule re-entered the atmosphere. What happened to all the other stuff?

The Apollo payload consisted of the Launch Escape System, the Apollo capsule, the service module, and the lunar lander. The launch escape system (LES) was located above the Apollo capsule and was jettisoned early in flight. The Lunar Excursion Module (LEM) was stored behind the service module. Once in Earth orbit, the capsule and Service Module were separated, the capsule rotated 180 degrees, and docked to the Lunar module. The lunar package was then separated from the third stage. The capsule, lander, and service module left Earth orbit heading for the moon, while the third stage was commanded into a solar orbit, to get it out of the way.

The Saturn-V was a three-stage, human-rated launch vehicle. Thirteen of the vehicles were launched. All three stages used the same fuel and oxidizer, liquid hydrogen and liquid oxygen (LOX). The first stage, S1C, had a dry weight of around 130 tons. It fell into the Atlantic Ocean when it had done its job. The second stage, about 40 tons dry, continued into Earth orbit with the crewed capsule. The third stage, S-IVB, was sent into a solar orbit. During the Apollo 13, 14,15, 16, and 17, the S-IVB stages were deliberately crashed into the Moon to perform seismic measurements used for characterizing the lunar interior.

The Apollo command module with the Astronauts re-entered the atmosphere, and landed in the Ocean. Most of these are now in Museums. The Service Modules, returning from the moon, were jettisoned before reentry and burned in the atmosphere.

The Lunar Excursion Modules got two Astronauts to the lunar surface, and later returned them to the orbiting Command Module. The LEM was made in two parts, and the lander, or lower, section remained on the Moon.
Most of the crewed portion of the LEM's were deliberately crashed into the lunar surface, to provide data for the

seismic instruments left on the surface. This includes the lunar modules, Eagle and Intrepid. The Lunar Module Aquarius, from Apollo-13, burned up in Earth's atmosphere in April, 1970. It had been the lifeboat for the astronauts, after the Service Module exploded on the way to the Moon. It was jettisoned before the manned capsule reentered the atmosphere. The Lunar Modules Antares, Falcon, and Challenger impacted the lunar surface.

SMM

The Solar Maximum Mission satellite, a GSFC mission, was designed to investigate Solar phenomena, particularly solar flares. It was launched on February 14, 1980.

In January 1981, three fuses in the SMM's attitude control system failed, causing it to rely on its magnetic torquers to maintain attitude. In this mode, only three of the seven instruments were usable, as the others required the satellite to be accurately pointed at the Sun. The use of the satellite's magnetic torquers prevented the satellite from being used in a stable position and caused it to "wobble" in its orbit.

Although not unique in this endeavor, the SMM was notable in that its useful lifetime was significantly increased by the direct intervention of a manned space repair mission. During STS-41-C in 1984, the Space Shuttle Challenger intercepted the SMM, maneuvering it into the shuttle's payload bay for maintenance and repairs. SMM had been fitted with a shuttle "grapple fixture" so that the shuttle's robot arm could grab it. During the mission, the SMM's entire attitude control system module and the electronics module for the an instrument were replaced, and a gas cover was placed over another instrument. SMM was the first on-orbit servicing mission in history.

The success of the SMM repair demonstrated beyond a doubt the feasibility of servicing a spacecraft in orbit, but at a high level of complexity, involving a Shuttle mission, and trained astronauts. These repairs were successfully completed, adding five years to the satellites working life. The spacecraft reentered the atmosphere and burned in December of 1989, taking some of the author's best flight software with it.

What is most interesting was an analysis of the modules brought back, after a 4-year period in orbit. Much was learned about the debris flux, and composition. This was from surfaces not specifically designed for particle collection, specifically, the thermal blankets. Analysis of these revealed valuable information on penetration depth, impact velocity, and particle size.

In addition, the returned hardware from the Hubble Space Telescope Servicing Missions provided new insight on space debris strikes. These included the large solar arrays. This evidence was used to validate debris models.

Space Station Mir

Quite a lot of lessons were learned from the impact of small debris on the Soviet Space Station Mir. The Environmental Effects payload was returned and studied, adding to knowledge applied to the ISS. It was in orbit for 3,644 days before reentering the atmosphere. In an amusing incident, a Kvant re-supply module was having trouble docking with the station. An EVA revealed a trash bag in the way. Seems you can't just leave the trash outside the door, in orbit, and have it picked up.

MIR was in a near-circular orbit of 354-375 kilometers. Atmospheric drag continually lowered the orbit, and it was re-boosted by the resupply vehicles. The station reentered and burned, landing in the South Pacific in March of 2001. The facility had been designed with a 5-year lifetime goal, but lasted 15.

Shuttle

Since the Shuttle orbiter returned to the ground and landed on a runway after its missions, it was carefully studied for the effects of collisions with orbital debris. This generally included chipping of the windows, and minor damage to the thermal tiles, which tended to contain the debris. In addition, the Shuttle's orbit during a mission was carefully inspected for detectable debris, and it could maneuver to prevent collisions. STS mission 48 in 1991 had the first avoidance maneuver, to miss Kosmos 955. Several other missions had avoidance maneuvers. On missions to the Space Station, the facility was approached so the Shuttle remained in the Stations debris shadow.

In 2006, a chunk of circuit board in orbit damaged one of Shuttle Atlantis' radiator panels. Later, a similar thing happened to Shuttle Endeavor.

In the 2003 Shuttle Columbia disaster, a large amount of debris hit the ground in a long path, as the Shuttle disintegrated during re-entry. Debris was strewn across hundreds of miles of Texas and Louisiana. The accident investigation board found that a hole had been punched in the leading edge of a wing, by a piece of insulating foam broken off the external fuel tank early in the ascent. There were 84,000 pieces of debris collected. This is stored at the Kennedy Space Center.

ISS

In normal operations, the Earth's magnetic field deflects charged particles from the station. Energetic space particles may pass through the station with negligible effect. Space debris is a problem, from discarded bolts to Zombie-Sats. These are all tracked, and on rare occasion, the station needs to do a damage avoidance maneuver to avoid a collision. To date, no evacuation of the Station has been necessary. It does a debris avoidance maneuver if there is a greater than 1 in 10,000 chance of a strike. This tends to happen about once per year.

It the warning comes too late for maneuvers (usually a few days), the crew shelters in the attached return capsule.

Trash comes down, with the simple yet costly expedient of having a logistics carrier burn up in the atmosphere on reentry.

The current ISS will reach end-of-life in the 2020's, and is too big to be allowed to re-enter in one piece. One or more follow-on stations will be built in orbit, re-using some modules from the ISS, and new modules launched from the ground. This is possible due to the modular nature of the ISS, and the lessons-learned during its on-orbit construction and use.

Russia currently has plans to remove some of its modules, and re-purpose them into a new facility, the Orbital Piloted Assembly and Experiment Complex (OPSEK). This is based on an estimated life on-orbit of 30 years, from the MIR experience.

The various nations that own parts of the station are responsible for their disposal. It the parts can not be re-purposed, they will be re-entered into the atmosphere in a controlled manner.

The LDEF Mission

The Long Duration Exposure Mission was placed in a 275 mile orbit by Space Shuttle Challenger in 1984. It spent 69 months in space, collecting information on long term exposure of materials to the space environment, and the meteoroid environment. It was to be periodically returned to Earth, refurbished, and re-orbited, but instead, the original mission was extended. It was recovered and returned to Earth by Shuttle Columbia. Although it was expected that meteoroid impacts would be the major phenomena, over 30% of the observed impacts were from debris. This lead to the realization that the debris environment was larger than suspected. In some cases, the impacting body was still embedded in LDEF's structure.

There was penetration of up to 40 mils of aluminum observed. In total, 34,000 impacts were cataloged, the largest being 0.57 cm.

Characterizing the "space dust" environment, it is estimated that over 14 million tons fall to the Earth's surface each years. The number of meteors entering the atmosphere is estimated to be between 10 and 20,000 tons. Most of these burn in the atmosphere, but when they do hit, they can be catastrophic. Generally, if the chunk is big enough to do damage, it can be tracked. Sometimes you miss them. Like in the Urals in Russia in February, 2013. This injured hundreds, but did not hit anything in the town. There could have been hundreds of deaths.

Saturn-Pegasus

The Pegasus payload was intended to supply much-needed information on the near-space meteoroid environment, that would influence the design of manned spacecraft. The main reason this information was vital was that space missions were lasting longer, and getting bigger. They were puncture targets for small chunks of rock, smaller than a grain of sand, traveling faster than a bullet. Armor was out of the question – it would add too much weight. What were the chances of a meteorite hit, and how damaging would it be? No one knew. The Saturn I missions SA-8, -9, and -10 were used to launch the three Pegasus spacecraft in 1965.

The Pegasus missions were considered by NASA in 1962, and contracted to Fairchild Hiller Corporation. Design and development took place at their facilities in Bladensburg and Rockville, MD, with assembly in Hagerstown, MD. Pegasus was a secondary payload on the Saturn vehicle, with the primary payload being a boilerplate Apollo spacecraft. The Apollo boilerplate acted as a payload fairing for the Pegasus spacecraft, which was stored inside what would have been the Apollo Service Module. The Pegasus was a 3,200 pound satellite in low Earth orbit, designed to study micro-meteoroid impacts, an area that was relatively unknown at the time. The satellite had large 95-foot wing panels that folded out from the satellite body, and included 116 detectors, a data and power system, and a telemetry and tracking system. A lot was learned from the missions besides the micro-meteoroid environment. The Pegasus provided information on the thermal effects of surface coating in space, the susceptibility of electronics to the radiation environment, the orbital thermal environment, flight dynamics characteristics, and many other factors.

The Pegasus was a very large and heavy payload, but, with the Saturn, size and weight were not of much concern. Folded in launch configuration, the Pegasus was 17 feet, 4 inches high, 7 feet wide, and 11 inches deep. It was constructed of aluminum alloy. The large deployed wings contained penetration surfaces for the measurement of micrometeorite impact. The wings had a total surface of 2,300 square feet.

The 11- by 16-inch wing panels were subdivided into 62 logic groups of from two to eight capacitors each. The capacitors are interconnected so that the satellite electronics package saw each logic group as one capacitor. A meteoroid hit on any panel was registered as a hit on the logic group in which that panel was located. Some capacitors on Pegasus I shorted in orbit, and it was necessary to remove logic groups; i.e., disconnect good or bad capacitors from the overall detection system. A new fusing arrangement was incorporated in the meteoroid detection system of Pegasus to fuse each capacitor individually. Before launch, the Pegasus satellites were known as A, B, and C.

A single malfunctioning capacitor left the other capacitors in the same logic group operating. When a malfunction occurred which was serious enough to warrant disconnection of the entire logic group, this could be done by ground command. The fusing arrangement worked successfully on Pegasus B and was installed on Pegasus C. The fuses could be blown by 50 milli-amps. The ground command to blow a capacitor fuse could "heal" the capacitor instead of blowing the fuse, depending upon the cause of the short. Each capacitor "healed" in this manner was a bonus benefit. Each time a capacitor was penetrated by a meteoroid, the material removed by the impact was vaporized, forming a conducting gas which discharged the capacitor. The gas, or plasma, dissipated almost immediately and the capacitor recharged within three one-thousandths of a second. The recharge event was what was recorded.

If seen on the screen of an oscilloscope, the "blip" caused by a penetration and momentary discharge of the capacitor would be a sharp saw-tooth below the horizontal line.

These blips were characteristic for each group of panels, providing a means of determining which group contained the penetrated panel.

When a panel was penetrated, several items of related information were recorded, including a cumulative count of hits classified according to panel thickness; an indication of the panel group penetrated; the attitude of the satellite with respect to both the Earth and the Sun, the temperature at various points on the spacecraft, the time at which each hit was recorded, and the condition of the power supply and other equipment supporting overall spacecraft operation.

Various levels of impact energy were differentiated through the use of panels of three different thicknesses. Directional information was gained by using a combined solar sensor-Earth sensor system. The aluminum sheets of the wing assemblies were separated by a layer of polymer plastic, forming a capacitor. It was charged with 40 volts. There are capacitor panels on both sides of the wing assembly, a total of 416 separate detector panels. In addition, the spacecraft determined its attitude with respect to the Earth and Sun, suing Earth and solar aspect sensors. A backup mode was spin-stabilization.

The Pegasus electronic system registered meteoroid penetrations of panel groups and stored a record of panel thickness, group number, and time of penetration. Pegasus attitude and certain temperatures were recorded on a timed schedule. The storage for the data was a 30,080 bit magnetic core memory unit. This could nominally hold the data for 6 to 8 hours of operation. The memory was read-out and transmitted 6 times, for redundancy, then cleared. This required about 1.5 minutes.

The Pegasus spacecraft was to detect meteoroids in the mass range of 10^{-7} to 10^{-4} grams, leading to an understanding of the on-orbit meteoroid environment in terms of density and direction. The measurements were taken between 500 and 800 kilometers. The detectors array used 116 capacitors of varying thicknesses over 185 square meters of area. Both real-time and stored data transmission were provided.

A new capacitor fusing arrangement was used on the second Pegasus after short circuits in the Pegasus I detection system hampered the capabilities of that satellite. This new system was working well after one month. It provided the ability to disconnect a single malfunctioning capacitor detector while leaving other capacitors in the same group of panels working. Thirty-six capacitors on Pegasus were found to be working improperly during the first four weeks and were disconnected by ground command to prevent a drain on the satellite's power supply. Of these 36 capacitors, four were isolated and disconnected.

Pegasus-2 returned data until August 29, 1968. It reentered the atmosphere on November 3, 1979. Another potential use of the third Pegasus experiment was announced in NASA Press Release 65-232 in July 1965. This was to attempt to return to Earth samples of the meteoroid punctured metal, with hopefully captured micrometeoroids. The Press Release states the purpose:

"This flight's primary purpose is to add information on the frequency of meteoroids to be encountered in near-Earth environment, for use in the design of future manned and unmanned spacecraft. The information was vitally needed with the increased emphasis on larger, long-life spacecraft, and the mission of the three-flight Pegasus program was to provide data necessary t o determine the magnitude of the meteoroid hazard."

"The engineering experiment consists of 4-8 aluminum sub-panels or "coupons" attached to Pegasus which could be quickly unhooked by an astronaut on an EVA and carried back to Earth in a Gemini or Apollo capsule. NASA officials emphasize that no decision has been made for an astronaut to rendezvous and retrieve the panels." Keep in mind, this was before any manned Apollo Capsule had flown.

"Although numerous experiments have been conducted in space, no materials punctured by meteoroids have been returned so far. Meteoroids are the countless small particles of matter flying in space at great speeds. When they enter the Earth's atmosphere, they burn -- as meteors - and those that reach the ground are known as meteorites."

At the time of the third Pegasus launch, the previous two Pegasus spacecraft were still on-orbit, returning data. Additional data on the micrometeorite environment was still coming in from Explorer XXIII, launched in 1964.

Skylab

After the Apollo program, with some spare Saturn's sitting around, the next project was the Skylab space station. This used a Saturn S-IVB upper stage as the structure for the station, launched by a Saturn-V with live first and second stages. The hydrogen fuel tank was re-purposed into the crewed facility. The payload to orbit was 170,000 pounds. The station was 82 feet long, 56 feet wide, and 36 feet in the other direction. It was quite visible from Earth, when the solar arrays caught the glint of the Sun. Astronauts were carried to the facility in-orbit on three missions in 1973-1974 by Apollo command and service modules launched on Saturn-Ib vehicles. A second Saturn-1b and Apollo stack was kept in standby in case a rescue mission was needed. There are no documented cases of problems with orbital debris, but the 170,000 pound Skylab itself posed a disposal problem

There were plans to use the Shuttle to repair and reboost Skylab, but the timing did not work out. Skylab was in orbit until 1979, when it reentered the atmosphere, splashing into the Pacific Ocean near Perth, Australia. A few pieces of debris were found on land.
Liquid waste was not recycled in Skylab as it is on the current ISS. Liquid and solid waste went into the large Oxygen tank in the facility. It burned when Skylab re-entered.

The German Space Agency, DLR, developed the designs for the DEOS (Deutsche Orbitale Servicing Mission) robotic spacecraft. It was supposed to be capable of capturing an "un-cooperative" target, which would then put into a "destructive re-entry". This is intended for tidying up orbital debris. Unfortunately, no funding has been forthcoming for this project.

NASA Orbital Debris Program Office

The NASA Orbital Debris Program Office is hosted at the Johnson Space Center, in Houston, Texas. Under their guidance, orbital debris models are developed and tested. Some models are used to predict future orbital debris environments. Orbital debris is tracked optically and via ground based radar. Examining spacecraft parts that have been returned from orbit by Shuttle-based servicing missions are very useful. The Shuttles themselves provided good examples of debris hits. Models are validated by JSC's Hypervelocity Impact Technology Facility.

The Inter-agency Space Debris Coordination Committee consists of members of 10 space-faring countries, and ESA, and addresses the growing problem.

The U. S.'s Orbital Debris Mitigation Standard Practices addresses the debris problem. It defines mitigation standard practices for satellites and upper stages to control release of particles greater than 5 mm, that will remain on-orbit for 25 years or more. A part of this is to assess the mission in terms of limiting the probability for explosion. This applies to all onboard sources of stored energy. Spacecraft are required to be designed such that collisions with debris smaller than 1 cm will not cause loss of control. Lethal, non-trackable debris is defined as that less than 10 cm. The U.S. Space Surveillance Network can track items down to 10 cm in LEO, or 1 meter in Geo.

End of life disposal is also addressed in three areas, atmospheric reentry, storage orbit, and retrieval. The reentry option assumes that the atmospheric drag will cause reentry within 25 years. This may involve the deployment of drag enhancement devices. Since even Cubesats are subject to this restriction, specific drag-enhancements for them have been developed. Reentry presents another problem, that of danger to humans and structures on the ground. The satellite must be re-entered in such a way that it enters the ocean in a remote location, with a 1 in 10,000 chance of human casualty. So far, so good.

A second option is maneuvering to a storage, or "graveyard" orbit. This doesn't solve the problem, it just postpones it. In the future, we can foresee an industry of spacecraft recycling in orbit. Between LEIO and medium orbits, the spacecraft is supposed to be placed with a perigee above 2,000 km, and an apogee below 19,700 km (below synchronous orbit). For medium altitude orbits, up to GEO, the storage orbit is a perigee above 20,700 km, and a apogee above 35,300 km. This will be below synchronous altitudes. For geosynchronous altitude, the graveyard orbit is 300 km above the geosynchronous altitude. If the spacecraft is in a heliocentric orbit (around the Sun), it can be maneuvered to collide with the Sun.

The third option is retrieval, which is difficult now with

the retirement of the Space Shuttle. That also makes the option of on-orbit repair moot at the moment, but NASA is actively exploring in-orbit robotic repair. Active removal involves more than technical issues. There are legal and ownership issues. On the high seas, it is the law that if you retrieve an un-crewed ship, it is yours. That's not true in space.

The RemoveDebris Project is addressing some possible solutions at the Surrey Space Centre ie U.K. It is going to the ISS, and deployed by the crew. It comes with its own debris – several smallsats. These will be released, and RemoveDebris will attempt to net them. It's main approach is a harpoon. The third part of the technology demonstration is that it will be sent into the atmosphere to burn up, using a 10 square meter drag chute. As opposed to most missions, reentry and burning are the goals.

The Remove Debris mission was deployed from the ISS in mod-September, 2018, and worked as expected. The net deployed properly. The Cubesat target expanded a balloon to increase its size. It is expected to reenter within a month. The next step is for the clean-up spacecraft to test out its harpoon with a flat target it will deploy, then deploy a sail, which will expedite reentry.

Airbus Industries is trying a variation of this approach for the elimination of Envisat Earth Observation Platform, a defunct ESA Mission launched in 2002. It is somewhat embarrassing to have the world's largest environmental satellite cluttering up orbit. It masses some 8 metric tones. A harpoon penetrator is fired with compressed gas, and easily penetrates the 3 cm composite honeycomb body panels. Then a set of spring loaded barbs extend, making the harpoon impossible to pull out. This harpoon is larger than the one considered for the RemoveDebris Project.

After the harpooned spacecraft is captured, the capturing satellite does a controlled reentry.

Tracking Junk, NORAD

Orbital Debris is tracked by radar. NORAD, the North American Aerospace Command, based in Colorado, tracks all detectable orbital entities, from large satellites to space junk, zombie-sats, and the larger pieces of debris, as well as near-Earth asteroids. The U. S. Space Surveillance Network can see objects 10 cm. or greater. They are deploying a new large telescope to improve their view of debris. This Space Surveillance Telescope will be able to see debris at Geosynchronous altitudes.

NORAD puts all this up on a website, in a standard format called the "two-line element" (TLE). This contains the Keplerian orbital elements, the set of data describing the orbit of anything around the Earth, for a given point in time (epoch). It is a legacy format form the 1960's, that still works. It includes two data items of 80 ASCII charters each (an IBM punch card format).

NORAD puts all this up on a website for your convenience, in a standard format called the "two-line element" (TLE). This contains the Keplerian orbital elements, the set of data describing the orbit of anything around the Earth, for a given point in time (epoch). It is a legacy format form the 1960's, that still works. It includes two data items of 80 ASCII charters each (an IBM punch card format).

Here is the format and contents of Line 1:.

Field	Columns	Content
1	1	Line number
2	3-7	Satellite number
3	8	Classification (U = unclassified)
4	10-11	Internat. Designator, last two digits of launch year
5	12-14	Launch number of teh year
6	15-17	Place of launch
7	19-20	Epoch, last two digits of year
8	21-32	Epoch, day of year, and fractional portion of day
9	34-43	First Time Derivative of the Mean Motion divided by two
10	45-52	Second Time Derivative of Mean Motion divided by six (decimal point assumed)
11	54-61	BSTAR drag term, (decimal point assumed)
12	63	number 0
13	65-68	element set number, incremented for new TLE
14	69	Checksum, modulo 19

Here is the format of line 2:

Field	Columns	Content
1	1	Line number

2	3-7	Satellite number
3	9-16	Inclination (degrees)
4	18-25	Right ascension of the ascending node (degrees)
5	27-33	Eccentricity (decimal point assumed)
6	35-42	Argument of perigee (degrees)
7	44-51	Mean Anomaly (degrees)
8	53-63	Mean Motion (revolutions per day)
9	64-68	Revolution number at epoch (revolutions)
10	69	Checksum (modulo 10)

You can also use this service: http://www.celestrak.com/NORAD/elements/

Need to watch out for space debris? Go here: http://satellitedebris.net/Database.

Wrap-up

We have made a mess in various Earth orbits. We have crashed spacecraft onto the moon, Venus, Mars, and Jupiter. At some point, we have to recycle, not dispose of. The ISS will be decommissioned, and quite a few parts can and will be used in follow-on facilities. Besides the concept of mining the asteroids, there is the feasibility of recycling old satellites.

The clean-up of the space debris problem is less technical than political, legal, and economic. There is no International agreement covering space debris, but the United Nations Committee on the Peaceful Uses of Space has issued guidelines. The U.S. is developing a "one-up, one-down" policy attached to its launch license procedure.

This involves the capture and disposal of a derelict satellite, after the primary mission is deployed.

The evolution of on-orbit robotic repair vehicles will lessen the number of derelict spacecraft in orbit by refueling and repairing them when possible, but could also attach a de-orbit rocket. A large number of projects for orbital debris mitigation have been studied, but these are expensive. We probably won't see any action until there is a "major event."

Bibliography

AIAA, *Orbital Debris Mitigation Techniques: Technical, Economic, and Legal Aspects,* 1992, ISBN-1563470233.

AIAA, *Special Project Report: Meo/Leo Constellations: U.S. Laws, Policies, and Regulations on Orbital Debris Mitigation,* 1999, ISBN-1563473518.

Albu-Schaffer, Alin "DLR's Robotic Technologies for Space Debris Mitigation and On-Orbit Servicing", Institute of Robotics and Mechatronics, (PowerPoint Presentation). Avail: http://www.unoosa.org/pdf/pres/stsc2013/2013iaf-05E.pdf

Allahdadi, Firooz (Ed) *Space debris detection and mitigation:,* 15-16 April 1993, Orlando, Florida (Proceedings / SPIE--the International Society for Optical Engineering), 1993, ISBN-0819411876.

Ashley, David M. *Risk Assessment of Space Debris Hazards for Global Positioning Spacecraft,* 1998, ASIN-B00IKVTUZ2.

Asimov, Isaac *Space Garbage,* 1989, ISBN-978-0440404446.

Baiocchib, Dave; Welser IV, William *Confronting Space Debris: Strategies and Warnings from Comparable Examples Including Deepwater Horizon,* 2010, ISBN-0833050567.

Baker, Howard *Space Debris:Legal and Policy Implications*, Springer, 1989, ISBN-0792301668.

Bell, Larry D. *Planetary Asteroid Defense Study: Assessing and Responding to the Natural Space Debris Threat*, 2012, ISBN-1288329687.

Bendisch, Joerg (ed) *Space Debris*, 2001, American Astronautical Society, ISBN-0877034737.

Bendisch, Joerg (ed) *Space Debris and Space Traffic Management 2003*, ISBN-0877035172.

Biesbroek, Robin, *Active Debris Removal in Space: How to Clean the Earth's Environment from Space Debris*, 2015, ISBN-1508529183.

Bohan, Humpherey M. *The Space Debris Problem and Solutions, Specifically the Disposal of the Centaur Rocket Body after Use*, 1998, ASIN-B00IKVUA6A.

Brower, Jared *Eliminating Space Debris: Applied Technology and Policy Prescriptions*, Fall 2007 - Project 07-02, 2008, ASIN-B00PV8XWUQ

Burke, John G. *Cosmic Debris: Meteorites in History*, University of California, ISBN-0520056515.

Campbell, J. W. *Project ORION orbital debris removal using ground-based sensors and lasers*, 1996, NASA/MSFC, ASIN- B00010TYEW.

Cerf, Max *Space Debris Cleaning Missions*, 2017, ISBN-3330877960.

Chan, F. Kenneth *Spacecraft Collision Probability*, AIAA, 2008 ISBN-1884989187.

Christiansen, E. L. *Solving Problems Caused by Small Micrometeoroid and Orbital Debris Impacts for Space-Walking Astronauts*, 2014, ASIN-B01D54FR96.

Chobotov, V. A. *Dynamics of Debris Motion and the Collision Hazard to Spacecraft Resulting from an Orbital Breakup*, 1988, ASIN-B00CQCTSQM.

Clifton, S.; Naumann, R. J. *Pegasus Satellite Measurements of Meteoroid Penetration*, February 16 - December 31, 1965, MSFC, Dec 1, 1966, Document ID 19670003479.

Datta; Lakshya Vaibhav; Guven, Ugur *Introduction to Space Debris: Challenges and Removal Techniques: Fundamentals of Space Debris Removal from Low Earth Orbit and Middle Earth Orbit,* 2013, ISBN-3659363405

De Andrade, Jr. Elias S*pace Debris: A Great Leap Forward We Won't Take*, ISBN-3330502169.

Jack Donahue, C*atastrophe on the Horizon: A Scenario-Based Future Effect of Orbital Space Debris,* 2012, ASIN-B012UQ2JQ4.

Eather, Robert H. *Space Debris Detection and Analysis,*

1996, ASIN-B00HCUUL28.

Edelstein, Karen S. "Orbital impacts and the space shuttle windshield," (SuDoc NAS 1.15:110594), 1995. avail: https://ntrs.nasa.gov/archive/nasa/casi.ntrs.nasa.gov/19950 019959.pdf

Esther, Elizabeth A. *RS-34 Phoenix In-Space Propulsion System Applied to Active Debris Removal Mission*, 2014, ASIN-B01ED7JJUM.

Fu, Wang Hai *Space Debris Introduction*, 1991, ISBN-7030234340.

Goldstein, Margaret J. *Garbage in Space: A Space Discovery Guide*, 2017, ISBN-1512425907.

Hildreth, Steven A. Threats to U.S. National Security Interests in Space: Orbital Debris Mitigation and Removal, 2014, ASIN-B00HTQH0DS.

Ireland, Susan *Dodging Bullets: The Threat of Space Debris to U.S. National Security*, 2012, ASIN-B012YT5KSG.

Johnson, Nicholas L.; McKnight Darren S. *Artificial Space Debris*, 1987, ISBN-0894640127.

Johnson, Nicholas L. *Orbital Debris Mitigation Requirements and the GRAIL Spacecraft*, 2014, ASIN-B01D54HI0C.

Johnson, Nicholas L. *Origin of the Inter-Agency Space Debris Coordination Committee*, 2014, ASIN-B01D54GIT4.

Katnik, Gregory A. D*ebris/Ice/Tps Assessment and Integrated Photographic Analysis of Shuttle Mission STS-93*, 2013, ISBN-1287277721.

Kendall, Kristen; *Space Shuttle Solid Rocket Booster Debris Assessment*, 2006, ASIN-B01H4EV17A.

Kessler, Donald J. *Space Debris, Asteroids and Satellite Orbits* (Advances in Space Research), 1985, ISBN-0080418422.

Kessler, Donald J. *Orbital Debris Environment for Spacecraft Designed to Operate in Low Earth Orbit*, 1989, ASIN-B00DD79L7K.

Klinkrad, Heiner *Space Debris: Models and Risk Analysis*, 2006, Springer, ISBN-354025448X .

J. C. Liou USA S*pace Debris Environment, Operations, and Measurement Updates*, 2015, ASIN-B01D54FD6I.

J. C. Liou, *Orbital Debris Challenges for Space Operations,* 2016, ASIN-B01D54FEFI.

Johnson, Nicholas, *History of On-orbit Satellite Fragmentations,* 14th edition, 2008, Orbital Debris Program Office, ISBN-1782661700.

Krisko, Paula H. *The Predicted Growth of the Low Earth Orbit Space Debris Environment: An Assessment of Future Risk for Spacecraft*, 2013, ISBN-1289284733.

Loftus, Joseph P. *Orbital Debris from Upper Stage Breakup*, 1989, AIAA, ISBN-0930403584.

Luu, Kim *Effects of Perturbations on Space Debris in Supersynchronous Storage Orbits* 1998, ASIN-B00IKVXJ4A.

Maethner, Scott R. *Space Debris Research Phase One Program: Abstracts from Published Papers* (1990-1993), 1994, ASIN-B00GBMIUP8.

Masevich, A.G. (Ed), T*he Problem of Space Pollutio*n, 1994, NASA/JSC, translated from the Russian, ASIN-B01LXLYVAD.

Owen, Ruth *Space Garbage* (Objects in Space), 2014, ISBN - 978-1477758663.

Long, Patrick *Space Junk Norms: US Advantages in Creating a Debris-Reducing Outer Space Norm*, 2012, ISBN-1249600499.

McNutt, Ross T. *Orbiting Space Debris: Dangers, Measurement and Mitigation*, 1992, ASIN-B00F0ZVULS.

Meshishnek, M. J. Overview of the Space Debris Environment, 1996, ASIN-B00GNNHGI2.

Milne, Anthony *Sky Static: The Space Debris Crisis*, 2002, ISBN-0275977498.

Nake, R. *Space Surveillance, Asteroids and Comets, and Space Debris*. Volume 3: Space Debris Summary Report, 1997, ASIN-B00I4EMCXC.

Page, Thornton & Page Lou Williams (ed) *Neighbours of the Earth. Planets, Comets, and the Debris of Space,* Macmillian, 1966, ASIN-B002SE359Y.

Paté-Cornell, M. Elisabeth *Assessment and management of the risks of debris hits during space station* EVAs (SuDoc NAS 1.26:112981), 1997, ASIN-B00010YRGW.

Pelton, Joseph N. N*ew Solutions for the Space Debris Problem*, 2015, Springer, ISBN-331917150X.

Pelton, Joseph N. Space Debris and Other Threats from Outer Space, 2013, Springer, ISBN-1461467136.

Powell, Jonathan *Cosmic Debris: What It Is and What We Can Do About It,* 2017, ISBN-3319510150.

Smirnow, Nickolay N. M. V. Lomonosov M. V. *Space Debris: Hazards Evaluation and Mitigation*, ISBN-9056993038 .

Stach Malgorzata, *Space Debris*, 2016, ISBN-10136507126X.

Stakem, Patrick H. *In-Space Robotic Repair and Servicing*

of Spacecraft, 2018, ASIN-B079C9BFQT.

Stakem, Patrick H. T*he Saturn Rocket and the Pegasus Missions*, 1965, 2013, PRRB Publishing, ASIN-B00BVA79ZW.

Stubbe, Peter State *Accountability for Space Debris: A Legal Study of Responsibility for Polluting the Space Environment and Liability for Damage Caused by Space Debris*, ISBN-9004314075.

Stuckey, W. K. *Lessons Learned From the Long Duration Exposure Facility*, 1993, Aerospace report TR-93(3935)-7. Avail: http://www.dtic.mil/dtic/tr/fulltext/u2/a266026.pdf

Swan, Cathy; Swan, Peter *Space Elevator Survivability Space Debris Mitigation*, 2015, ISBN- 1329093925.

Vasile, Massimiliano and Minisci, Edmondo *Asteroid and Space Debris Manipulation*, 2016, AIAA, ISBN-1624103235.

Walz-Chojnacki, Greg; Reddy, Frank *Pollution in Spac*e (Isaac Asimov's New Library of the Universe), 1995, ISBN - 978-0836811964.

Williams, Brando *Analyzing the Space Launch System Debris Environment*, 2015, ASIN-B01BB1ESRC.

Williamsen, J. E. *Vulnerability of Manned Spacecraft to Crew Loss from Orbital Debris Penetration,* 2015, NASA Marshall Space Flight Center, ASIN-B015AL6DWC.

Yang, Leping; Zhang, Qingbin *Dynamics and Design of Space Nets for Orbital Capture*, 2017, ISBN-3662540622.

Zedd, Michael *Concept of Operations for the Dust Dispenser Spacecraft for Active Orbital Debris Removal*, 2014, ASIN-B00VCQGOQ0.

Resources

NASA Orbital Debris Program Office; http://orbitaldebris.jsc.nasa.gov

Final Report, Pegasus Program, October 1965, Fairchild Hiller, Space Systems Division. NTIS N68-8704.

NASA Tech Reports Library, http://ntrs.nasa.gov

"The GEO Graveyard May Not Be Permanent," Staff, Tech Space, Nov. 2010, www.spacedaily.com

National Research Council, Division on Engineering and Physical Sciences, *Orbital Debris: A Technical Assessment*, 1995, ISBN-1558671587.

National Research Council and Division on Engineering and Physical Sciences, *Limiting Future Collision Risk to Spacecraft: An Assessment of NASA's Meteoroid and Orbital Debris Programs*, 2011, ISBN-0309219744.

National Research Council and Division on Engineering and Physical Sciences, *Summary of the Workshop to*

Identify Gaps and Possible Directions for NASA's Meteoroid and Orbital Debris Programs, ISBN-0309215153.

National Research Council, *Limiting Future Collision Risk to Spacecraft: An Assessment of NASA's Meteoroid and Orbital Debris Programs by Committee for the Assessment of NASA's Orbital Debris Programs*, 2011, ASIN-B01FIXY2UW.

*Protecting the Space Station from Meteoroids and Orbital Debris*P, 1997, National Academies Press, ISBN-0309056306.

United States. Congress, House. Committee on Science, Space, and Technology. Subcommittee on Space Science and Applications, Orbital space debris hearing before the Subcommittee on Space Science and Applications of the Committee on Science, Space, and Technology, House of Representatives, 1988, avail: https://catalog.hathitrust.org/Record/008516148

Protecting the Space Shuttle from Meteoroids and Orbital Debris, ISBN-058500207X.

World Spaceflight News, 2000 Earth Orbital Debris - NASA Research on Satellite and Spacecraft Effects, CD, 2000, ISBN-1893472280.

MacLay, Timothy D.; McKnight, Darren S. "Space Environmental, Legal and Safety Issues," 1995, Orlando, Florida, Proceedings, SPIE, Apr 1995, SPIE, ISBN-

0819418366.

NASA, "Vulnerability of Space Station Freedom modules a study of the effects of module perforation on crew & equipment : final report", contract NCC8-28, ... 1993-14 March 1995 (SuDoc NAS 1.26:205031) ASIN-B000111IZE.

U. S. Government Accountability Office, *Space Program: Space Debris a Potential Threat to Space Station and Shuttle*: Imtec-90-18, 2013, ISBN-1289291012.

U. S. Government Accountability Office, *Space Station: Delays in Dealing with Space Debris May Reduce Safety and Increase Costs*: Imtec-92-50, 2013, ISBN-1287291392.

Orbital Debris: *A Technical Assessment by Commission on Engineering and Technical Systems, 1995,* The Commission on Engineering and Technical Systems; Committee on Space Debris; National Research Council; Division on Engineering and Physical Sciences, ASIN-B01A64BP9E.

National Security Council, Inter-agency Group (Space) Report on Orbital Debris for National Security Council, Washington, D.C., 1989, ASIN-B01M16PPAJ.

"Proceedings of the First European Conference on Space Debris" :Darmstadt, Germany, 5-7 April 1993, European Conference on Space Debris Darmstadt, Germany), ASIN-B007HDY8EM.

"Proceedings of the Second European Conference on Space Debris: ESOC, Darmstadt, Germany, 17-19 March 1997, ISBN-9290922559.

"Technical report on space debris: Text of the report adopted by the Scientific and Technical Subcommittee of the United Nations Committee on the Peaceful Uses of Outer Space, 1999, ISBN-9211008131.

U.S. General Accounting Office, "Space program space debris a potential threat to space station and Shuttle" report to the Chairman, Committee on Science, Space, and Technology, House of Representatives (SuDoc GA 1.13:IMTEC-90-18), ASIN-B000107IR2.

Office of Technology Assessment, "ORBITING DEBRIS: A SPACE ENVIRONMENTAL PROBLEM: BACKGROUND PAPER," 1990, ASIN-B000LY9B6O.

The United Nations Office for Outer Space Affairs, *Space Debris, and Spy Satellites*, 2012, ASIN-B007S7TW84.

SCIENTIFIC ADVISORY BOARD (AIR FORCE) WASHINGTON DC, *Report on Space Surveillance, Asteroids and Comets, and Space Debris,* Volume 1: Space Surveillance, 1997.

Advances In Space Research Volume 19, Number 2, 1997, *Space Debris,* 1997, ISBN-B000HKGVF6.

NASA, *Joint Polar Satellite System (JPSS) micrometeoroid and orbital debris (MMOD) assessment,*

2017, ASIN-B01NBYH6AQ.

U.S. Government *Orbital Debris Standard Practices,* avail:https://orbitaldebris.jsc.nasa.gov/library/usg_od_standard_practices.pdf

NASA, JSC *Orbital Debris Quarterly News*, avail: https://www.orbitaldebris.jsc.nasa.gov/quarterlynews/subscription.html

https://www.orbitaldebris.jsc.nasa.gov/quarterly-news/pdfs/odqn-articles-index/odqn-quarterly-articles_dec%202016.pdf

Muth, Joseph *A. Dust Collection on Servicable Satellites*, N86-30599 (SMM Parts return) avail: https://ntrs.nasa.gov/archive/nasa/casi.ntrs.nasa.gov/19860021127.pdf

https://www.orbitaldebris.jsc.nasa.gov/library/un_report_on_space_debris99.pdf

https://www.orbitaldebris.jsc.nasa.gov/library/usg_od_standard_practices.pdf

https://www.orbitaldebris.jsc.nasa.gov/library/satellitefraghistory/tm-2008-214779.pdf

https://www.orbitaldebris.jsc.nasa.gov/library/a-technical assessment.pdf

https://www.orbitaldebris.jsc.nasa.gov/library/a-technical-assessment.pdf

https://www.orbitaldebris.jsc.nasa.gov/library/iadc_mitigation_guidelines_rev_1_sep07.pdf

https://www.orbitaldebris.jsc.nasa.gov/library/space-debris-mitigation-guidelines_copuos.pdf

https://nodis3.gsfc.nasa.gov/displayDir.cfm?t=NPR&c=8715&s=6A

https://standards.nasa.gov/standard/nasa/nasa-hdbk-871914

NASA Technical Reports Server, https://www.sti.nasa.gov/

Gravity, 2013 film,
 https://www.netflix.com/bw/title/70274337.

Space Debris 101, The Aerospace Corporation
https://aerospace.org, article = space-debris-101

The NASA Orbital Debris Program Office, located at the Johnson Space Center, is recognized world-wide for its initiative in addressing orbital debris.

https://www.iarpa.gov/newsroom/article/orbital-debris-detection-and-tracking-rfi

wikipedia, various

Glossary

AIAA – American Institute of Aeronautics and Astronautics.
Apogee – altitude of an orbits closest approach to Earth.
AR&D – Autonomous Rendezvous and Docking.
ASAT – anti-satellite (weapon).
ASIN – Amazon Standard Inventory Number
ATP – authority to proceed.
B612 Foundation – Formed to protect Earth from asteroid impacts.
BAA – Broad Agency Announcement (U. S. Government)
BD – Ballistic Missile Defense
COPUOS – (U.N.) Committee on the Peaceful Uses of Outer Space.
CPU – central processing unit

CRADA – Cooperative Research and Development Agreement (U. S. Government and industry)
CSA – Canadian Space Agency.
DARPA – (U. S.) Defense Advanced Research Projects Agency.
DLR – German Space Agency (Deutsches Zentrum für Luft- und Raumfahrt)
DOF – degrees of freedom
ELV – Expendable Launch Vehicle.
ESA – European Space Agency
EVA – Extra Vehicular Activity- involving an Astronaut with suit and maneuvering unit in space.
FAR – (US) Federal Acquisition Regulations

GEO – geosynchronous Earth orbit, 22,236 miles.
GHz – giga (109) hertz.
Giga - 109
GOES – NASA/NOAA Geostationary Operational Environmental Satellite
Graveyard orbit – a place to park end-of-life satellites.
Gray - unit of radiation, =100 rad
GSFC – Goddard Space Flight Center, Greenbelt, Maryland. NASA Center for unmanned spacecraft near Earth.
HIT-F (NASA-JSC) Hypervelocity Impact Technology Facility.
IADC – Interagency Space Debris Coordination Committee
IDD – Interface Definition Document.
Intelsat - International Telecommunications Satellite Organization.
IP – Intellectual Property
ISBN – International Standard Book Number.
ISS – International Space Station.
JAXA - Japan Aerospace Exploration Agency
JSC – NASA Johnson Space Center, Texas.
JSOC – U. S. Joint Space Operations Center
Kessler Syndrome – a cascade effect, where debris multiplies by collisions.
LDEF – Long Duration Exposure Facility
LEO – Low Earth Orbit
LIDS – Low impact docking system.
LSP – NASA launch services program.
LTG – LEO to GEO.
LV – launch vehicle.
MDA – (U. S.) Missile Defense Agency

MEO – medium earth orbit, above 2,000 km, below geostationary.
MES – mission extension services,
MEV-1 (Orbital-ATK) Mission Extension Vehicle-1.
mil – one thousandth of an inch; 0.0254 millimeter.
MMOD - micrometeoroid and orbital debris
MSFC – Marshall Space flight Center, Huntsville, Alabama.
NASA – National Aeronautics and Space Administration (USA)
NIST – National Institutes of Standards and Technology.
NOAA – National Oceanographic and Atmospheric Administration. (USA)
NORAD – North American Air Defense command
NAE – (U.S.) National Academy of Engineering.
NAS – (U. S.) National Academy of Science.
NRC – (U.S.) National Research Council.
NRL – U.S. Naval Research Center.
ODPO – (NASA) Orbital Debris Program Office
ODQN – Orbital Debris Quarterly News, from NASA, JSC.
OLEV - Orbital Life Extension Vehicles
OSSL - Orbital Satellite Services, LTD.
ORU – Orbital Replacement Unit.
ProxOps – proximity operations.
Perigee – point in an orbit furthest from the Earth.
POES – Polar orbiting environmental satellite.
RCS – reaction control system
RFI – Request for Information; radio frequency interference.
RNS – Relative Navigation System
RSGS – Robotic servicing of geosynchronous satellites.

SCM – Soft Capture Mechanism.
SIS – Space Infrastructure Servicing
Socrates – Satellite Orbit Conjunction Reports Assessing Threatening Encounters in Space.
SPIE-- International Society for Optical Engineering.
SSCO – Satellite Servicing Capabilities Office, NASA, GSFC.
SSN - Space Surveillance Network.
SSTL – Surrey Satellite Technology Ltd. (U.K.)
STS – Space Transportation System (USA) Shuttle.
TDRS – Tracking and Data Relay Satellite.
TLE – two line elements
ULA – United Launch Alliance, commercial launch services company.
Ullage – unusable fuel left in an "empty" tank.
USAF – United States Air Force.
Whipple shield - multi-layered shield designed so that the first layer breaks up the impacting object; the second layer breaks those fragments into smaller objects, and so on until the fragments are too small to penetrate the last layer.
Zombie-Sat – dead satellite posing a danger to other s pacecraft

If you enjoyed this book, you may have an interest in some of the author's other books:

Stakem, Patrick H. *Floating Point Computation*, 2013, PRRB Publishing, ISBN-152021619X.

Stakem, Patrick H. *Architecture of Massively Parallel Microprocessor Systems*, 2011, PRRB Publishing, ISBN-1520250061.

Stakem, Patrick H. *Multicore Computer Architecture*, 2014, PRRB Publishing, ISBN-1520241372.

Stakem, Patrick H. *Personal Robots*, 2014, PRRB Publishing, ISBN-1520216254.

Stakem, Patrick H. *RISC Microprocessors, History and Overview*, 2013, PRRB Publishing, ISBN-1520216289.

Stakem, Patrick H. *Robots and Telerobots in Space Application*s, 2011, PRRB Publishing, ISBN-1520210361.

Stakem, Patrick H. *The Saturn Rocket and the Pegasus Missions, 1965,* 2013, PRRB Publishing, ISBN-1520209916.

Stakem, Patrick H. *Visiting the NASA Centers, and Locations of Historic Rockets & Spacecraft,* 2017, PRRB Publishing, ISBN-1549651205.

Stakem, Patrick H. *Microprocessors in Space*, 2011, PRRB Publishing, ISBN-1520216343.

Stakem, Patrick H. Computer *Virtualization and the Cloud*, 2013, PRRB Publishing, ISBN-152021636X.

Stakem, Patrick H. *What's the Worst That Could Happen? Bad Assumptions, Ignorance, Failures and Screw-ups in Engineering Projects, 2014,* PRRB Publishing, ISBN-1520207166.

Stakem, Patrick H. *Computer Architecture & Programming of the Intel x86 Family, 2013,* PRRB Publishing, ISBN-1520263724.

Stakem, Patrick H. *The Hardware and Software Architecture of the Transputer*, 2011, PRRB Publishing, ISBN-152020681X.

Stakem, Patrick H. *Mainframes, Computing on Big Iron*, 2015, PRRB Publishing, ISBN- 1520216459.

Stakem, Patrick H. *Spacecraft Control Centers*, 2015, PRRB Publishing, ISBN-1520200617.

Stakem, Patrick H. *Embedded in Space,* 2015, PRRB Publishing, ISBN-1520215916.

Stakem, Patrick H. *A Practitioner's Guide to RISC Microprocessor Architecture*, Wiley-Interscience, 1996, ISBN-0471130184.

Stakem, Patrick H. *Cubesat Engineering*, PRRB Publishing, 2017, ISBN-1520754019.

Stakem, Patrick H. *Cubesat Operations*, PRRB Publishing, 2017, ISBN-152076717X.

Stakem, Patrick H. *Interplanetary Cubesats*, PRRB Publishing, 2017, ISBN-1520766173 .

Stakem, Patrick H. Cubesat Constellations, Clusters, and Swarms, Stakem, PRRB Publishing, 2017, ISBN-1520767544.

Stakem, Patrick H. *Graphics Processing Units, an overview*, 2017, PRRB Publishing, ISBN-1520879695.

Stakem, Patrick H. *Intel Embedded and the Arduino-101, 2017,* PRRB Publishing, ISBN-1520879296.

Stakem, Patrick H. *Orbital Debris, the problem and the mitigation*, 2018, PRRB Publishing, ISBN-*1980466483*.

Stakem, Patrick H. *Manufacturing in Space*, 2018, PRRB Publishing, ISBN-1977076041.

Stakem, Patrick H. *NASA's Ships and Planes*, 2018, PRRB Publishing, ISBN-1977076823.

Stakem, Patrick H. *Space Tourism*, 2018, PRRB Publishing, ISBN-1977073506.

Stakem, Patrick H. *STEM – Data Storage and*

Communications, 2018, PRRB Publishing, ISBN-1977073115.

Stakem, Patrick H. *In-Space Robotic Repair and Servicing*, 2018, PRRB Publishing, ISBN-1980478236.

Stakem, Patrick H. *Introducing Weather in the pre-K to 12 Curricula, A Resource Guide for Educators*, 2017, PRRB Publishing, ISBN-1980638241.

Stakem, Patrick H. *Introducing Astronomy in the pre-K to 12 Curricula, A Resource Guide for Educators*, 2017, PRRB Publishing, ISBN-198104065X.
Also available in a Brazilian Portuguese edition, ISBN-1983106127.

Stakem, Patrick H. *Deep Space Gateways, the Moon and Beyond*, 2017, PRRB Publishing, ISBN-1973465701.

Stakem, Patrick H. *Exploration of the Gas Giants, Space Missions to Jupiter, Saturn, Uranus, and Neptune*, PRRB Publishing, 2018, ISBN-9781717814500.

Stakem, Patrick H. *Crewed Spacecraft*, 2017, PRRB Publishing, ISBN-1549992406.

Stakem, Patrick H. *Rocketplanes to Space*, 2017, PRRB Publishing, ISBN-1549992589.

Stakem, Patrick H. *Crewed Space Stations*, 2017, PRRB Publishing, ISBN-1549992228.

Stakem, Patrick H. *Enviro-bots for STEM: Using Robotics in the pre-K to 12 Curricula, A Resource Guide for Educators,* 2017, PRRB Publishing, ISBN-1549656619.

Stakem, Patrick H. *STEM-Sat, Using Cubesats in the pre-K to 12 Curricula, A Resource Guide for Educators*, 2017, ISBN-1549656376.

Stakem, Patrick H. *Lunar Orbital Platform-Gateway,* 2018, PRRB Publishing, ISBN-1980498628.

Stakem, Patrick H. *Embedded GPU's*, 2018, PRRB Publishing, ISBN- 1980476497.

Stakem, Patrick H. *Mobile Cloud Robotics*, 2018, PRRB Publishing, ISBN- 1980488088.

Stakem, Patrick H. *Extreme Environment Embedded Systems,* 2017, PRRB Publishing, ISBN-1520215967.

Stakem, Patrick H. *What's the Worst, Volume-2*, 2018, ISBN-1981005579.

Stakem, Patrick H., *Spaceports*, 2018, ISBN-1981022287.

Stakem, Patrick H., *Space Launch Vehicles*, 2018, ISBN-1983071773.

Stakem, Patrick H. *Mars*, 2018, ISBN-1983116902.

Stakem, Patrick H. *X-86, 40th Anniversary ed*, 2018, ISBN-1983189405.

Stakem, Patrick H. *Lunar Orbital Platform-Gateway*, 2018, PRRB Publishing, ISBN-1980498628.

Stakem, Patrick H. *Space Weather*, 2018, ISBN-1723904023.

Stakem, Patrick H. *STEM-Engineering Process*, 2017, ISBN-1983196517.

Stakem, Patrick H. *Space Telescopes,* 2018, PRRB Publishing, ISBN-1728728568.

Stakem, Patrick H. *Exoplanets*, 2018, PRRB Publishing, ISBN-9781731385055.

Stakem, Patrick H. *Planetary Defense*, 2018, PRRB Publishing, ISBN-9781731001207.

Patrick H. Stakem *Exploration of the Asteroid Belt*, 2018, PRRB Publishing, ISBN-1731049846.

Patrick H. Stakem *Terraforming*, 2018, PRRB Publishing, ISBN-1790308100.

Patrick H. Stakem, *Martian Railroad,* 2019, PRRB Publishing, ISBN-1794488243.

Patrick H. Stakem, *Exoplanets,* 2019, PRRB Publishing, ISBN-1731385056.

Patrick H. Stakem, *Exploiting the Moon,* 2019, PRRB

Publishing, ISBN-1091057850.

Patrick H. Stakem, *RISC-V, an Open Source Solution for Space Flight Computers,* 2019, PRRB Publishing, ISBN-1796434388.

Patrick H. Stakem, *Arm in Space*, 2019, PRRB Publishing, ISBN-9781099789137.

Patrick H. Stakem, *Extraterrestrial Life*, 2019, PRRB Publishing, ISBN-978-1072072188.

Patrick H. Stakem, *Space Command*, 2019, PRRB Publishing, ISBN-978-1693005398.

CubeRovers, A Synergy of Technologys, 2020, PRRB Publishing, ISBN-979-8651773138.

Robotic Exploration of the Icy moons of the Gas Giants. 2020, PRRB Publishing, ISBN- 979-8621431006

Hacking Cubesats, 2020, PRRB Publishing, ISBN-979-8623458964.

History & Future of Cubesats, PRRB Publishing, ISBN-979-8649179386.

Hacking Cubesats, Cybersecurity in Space, 2020, PRRB Publishing, ISBN-979-8623458964.

Powerships, Powerbarges, Floating Wind Farms: electricity when and where you need it, 2021, PRRB

Publishing, ISBN-979-8716199477.

Hospital Ships, Trains, and Aircraft, 2020, PRRB Publishing, ISBN-979-8642944349.

<u>2020/2021 Releases</u>

CubeRovers, a Synergy of Technologys, 2020, ISBN-979-8651773138

Exploration of Lunar & Martian Lava Tubes by Cube-X, ISBN-979-8621435325.

Robotic Exploration of the Icy moons of the Gas Giants, ISBN- 979-8621431006.

History & Future of Cubesats, ISBN-978-1986536356.

Robotic Exploration of the Icy Moons of the Ice Giants, by Swarms of Cubesats, ISBN-979-8621431006.

Swarm Robotics, ISBN-979-8534505948.

Introduction to Electric Power Systems, ISBN-979-8519208727.

Centros de Control: Operaciones en Satélites del Estándar CubeSat (Spanish Edition), 2021, ISBN-979-8510113068.

Exploration of Venus, 2022, ISBN-979-8484416110.

Patrick H. Stakem, *The Search for Extraterrestial Life,* 2019, PRRB Publishing, ISBN-1072072181.

The Artemis Missions, Return to the Moon, and on to Mars, 2021, ISBN-979-8490532361.

James Webb Space Telescope. A New Era in Astronomy, 2021, ISBN-979-8773857969.

www.ingramcontent.com/pod-product-compliance
Lightning Source LLC
Chambersburg PA
CBHW030508220526
45464CB00006B/2714